MATHEMATICAL CHALLENGES FOR THE MIDDLE GRADES

FROM THE

ARITHMETIC TEACHER

Compiled and edited
by

William D. Jamski
Indiana University Southeast
New Albany, Indiana

NATIONAL COUNCIL OF TEACHERS OF MATHEMATICS

Library of Congress Cataloging-in-Publication Data:

Mathematical challenges for the middle grades : from the Arithmetic teacher / compiled and edited by William D. Jamski.
p. cm.
Includes bibliographical references.
ISBN 0-87353-296-1
1. Mathematics—Problems, exercises, etc. I. Jamski, William D. II. Arithmetic teacher.
QA43.M299 1990
510'.76—dc20 90-37386
CIP

Printed in the United States of America

CONTENTS

Preface v

Numbers (Problems 1–9) 1

Computation (Problems 10–37) 3

"Oldies But Goodies" (Problems 38–66) 10

Special Counting (Problems 67–77) 16

Offbeat and Unusual (Problems 78–87) 19

Geometry (Problems 88–118) 21

Probability (Problems 119–125) 30

Answers/References 32

PREFACE

This book is intended for, and dedicated to, those middle-grade mathematics teachers who are constantly searching for a wide variety of appropriately challenging and motivating problems with which to supplement regular textbook offerings for their students in grades five through eight. The 125 problems presented here, whether in form or concept, were found in the *Arithmetic Teacher* over the last fifteen years. References are given both to the articles that contained these problems as well as to related articles. Teachers should consult these works if they wish to explore fuller development of, or alternative approaches to, the ideas.

As the editor of the collection, I intend this collection of problems to accomplish at least three things: (1) provide a source of supplementary mathematical items and activities for middle-grade students; (2) recognize the continuing role of the *Arithmetic Teacher* in emphasizing the importance of problem solving; and (3) acknowledge the contribution of the individual authors, whose ideas I have tried to present while taking only minor editorial liberties.

NUMBERS

1. Mysterious Number

Pick a three-digit number in which the digit in the hundreds place is larger than the digit in the ones place. Reverse the number and subtract it from the original. If necessary, add lead zeros to the answer to make a three-digit number. Reverse the answer and add that number to the answer. What do you get? (Try a few examples.)

2. Perfect Numbers

A perfect number is one whose proper factors, not including the number itself, sum to the given number. Examples are $6 = 1 + 2 + 3$ and $28 = 1 + 2 + 4 + 7 + 14$. Show that 496 is a perfect number.

3. Primed for Action

How many factors do prime numbers have? How would you describe the number of factors that perfect squares have?

4. These Numbers Are Friendly

Two numbers are friendly (or amicable) if each is the sum of the proper divisors of the other. Show that 284 and 220 are friendly.

5. Fibonacci Numbers Don't Fib

What are the next three numbers in this sequence? 1, 1, 2, 3, 5, 8, ?, ?, ? (These numbers are named after a famous Italian mathematician of the Middle Ages.)

6. Palindromes Are Your Pals

Palindromes are numbers, such as 66, 343, and 2662, that can be read the same both forward and backward. Many nonpalindrome numbers can be converted into palindromes by reversing the digits of the number and adding the result to the original, continuing the process until a palindrome is achieved. (Note: Some numbers have been processed a large number of times without becoming palindromes.) How many steps are needed to convert these numbers into palindromes: (a) 32, (b) 458, (c) 748?

7. Look Out for Hailstone Numbers

The hailstone procedure starts with a counting number. If the number is even, divide it by 2; if it is odd, multiply it by 3 and add 1. Continue this process until you reach 1. As examples, 4 goes to 2 and then 1, taking two steps, and 3 ($3 \rightarrow 10 \rightarrow 5 \rightarrow 16 \rightarrow 8 \rightarrow 4 \rightarrow 2 \rightarrow 1$) takes seven steps. Of the numbers 5 to 10, which one shows the hailstone effect best by taking the most steps to reach 1?

8. Triangular Numbers

Given that $1 = 1$, $1 + 2 = 3$, $1 + 2 + 3 = 6$, and $1 + 2 + 3 + 4 = 10$, what does $1 + 2 + 3 + 4 + 5 + 6 + \ldots + n$ equal?

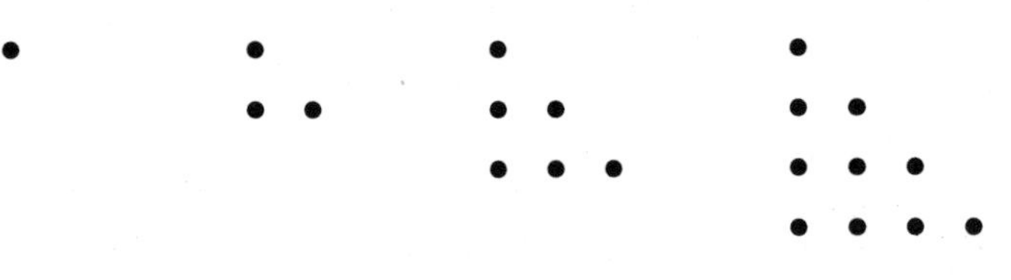

9. Don't Get Buried under Snowflake Numbers

The first five hexagonal snowflakes are diagrammed below. Without further drawing, determine how many dots will be in the next two snowflakes.

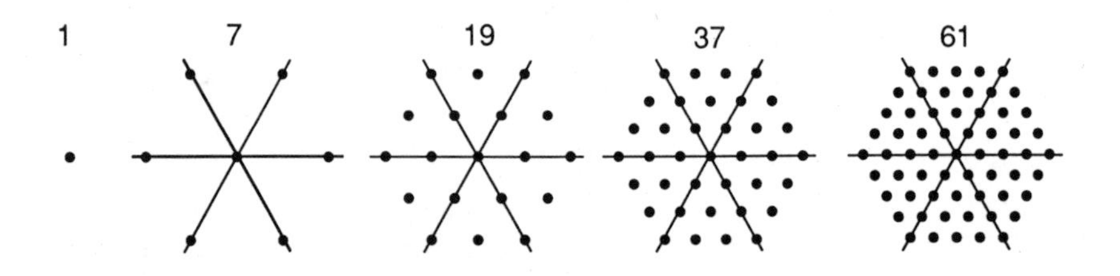

COMPUTATION

10. Remember the Bicentennial

Arrange the numbers 74, 148, 222, 296, 370, 444, 592, 666, 740, and 888 in the circles so that the sums along each side of this star are 1776.

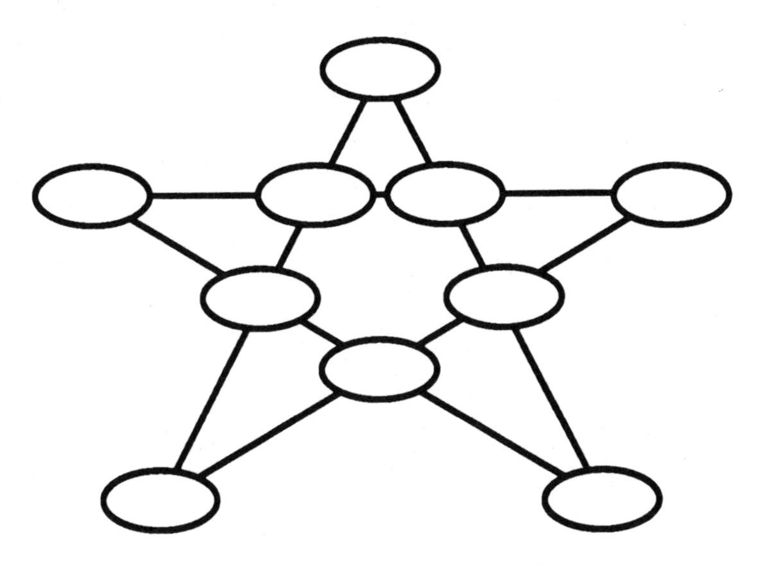

11. It's Magic

In this example, the numbers 1 to 9 are placed so that each side of the triangle sums to 17. Using the same numbers, find any other magic triangles whose sides sum to between 17 and 24.

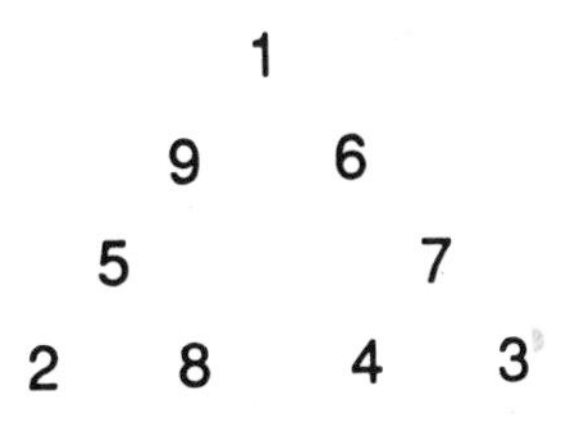

12. Hexagons Can Be Magic, Too

Using the numbers 1 to 19, complete this magic hexagon in which all diagonals and columns must sum to the same number.

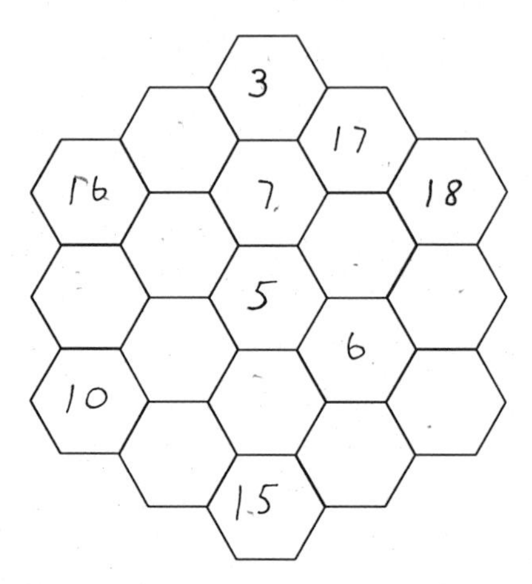

13. And a Partridge in a Pear Tree

In "The Twelve Days of Christmas," how many gifts in all did "my true love give to me"?

14. Clear the Table

Complete the following tables:

+		27	
46	64		
79			134
		60	

×	23		85
83			
	1288	2464	
			8415

15. Mirror, Mirror

Pick any three-digit number, for example, 762. Take its "mirror number," in this case, 267. Find the difference between the numbers. Continue the process until 0 results. What is the largest number of steps needed?

$$\begin{array}{r} 762 \\ -267 \\ \hline 495 \end{array} \qquad \begin{array}{r} 594 \\ -495 \\ \hline 99 \end{array} \qquad \begin{array}{r} 99 \\ -99 \\ \hline 0 \end{array}$$

16. Mining Properties

List all the properties that you can find in this multiple triangle:

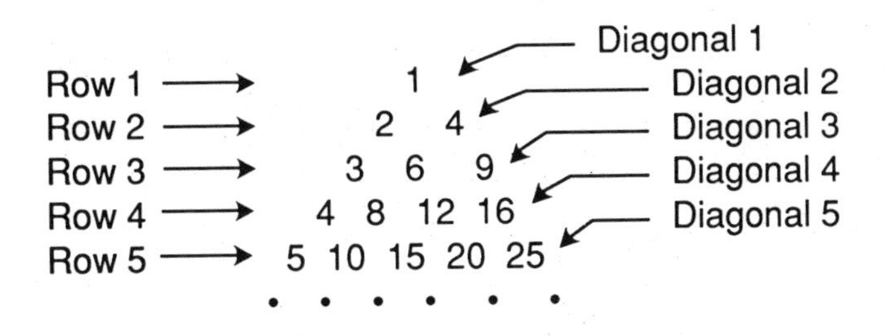

17. Multiplication Checks

A student computed 13×27 as below. How would the same student do 19×32?

1	27√
2	54
4	108√
8	216√
	351

18. 'Mazing Multiplication

Find the entrance and the path to reach the exit product. No diagonal moves are allowed. All numbers crossed must be multiplied. (A factor tree might help.)

8	5	11	Exit
2	13	3	3927
11	17	7	

19. Crazy Eights

Choose four consecutive even (or odd) counting numbers. Take the product of the middle two and subtract the product of the first and the last. Try a few samples and formulate a rule.

20. Reverse Gears

When the digits in this multiplication problem are reversed, the product is the same. Find the pattern for which this procedure works.

$$\begin{array}{r} 36 \\ \times 42 \\ \hline 1512 \end{array} \qquad \begin{array}{r} 63 \\ \times 24 \\ \hline 1512 \end{array}$$

21. Stretch 'em and Shrink 'em

Place the numbers 6, 2, 4, 3, and 8 in the boxes below to make the largest possible product. Repeat to make the smallest possible product.

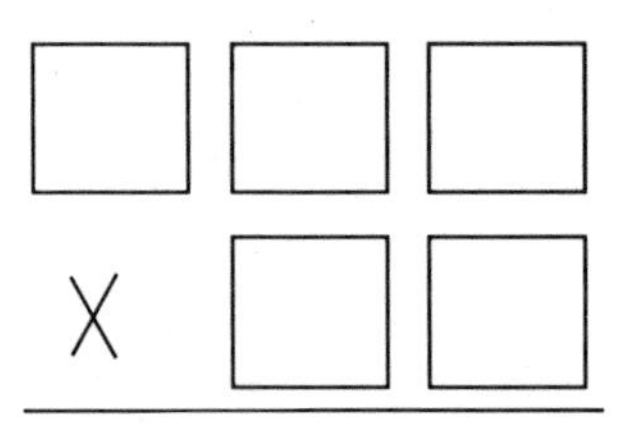

22. Lost Your Page?

You lost your page when you closed a book, but you remembered that the product of the pages showing was 16 002. What are the page numbers?

23. Alphanumerics

Using A = 1, B = 2, C = 3, . . ., Z = 26, find the word in which the product of the letter values comes as close to 1 000 000 as possible. (As an example, "CAT" comes to $3 \times 1 \times 20 = 60$.)

24. Producing Product Dates

March 30, 1990 is the first "product date" of the 1990s, since $3 \times 30 = 90$. How many more product dates are there in the 1990s?

25. Double Your Money, Double Your Fun

A rich neighbor offers Samantha a choice of $600 for a 16-day house-painting job or 1 cent the first day, twice as much the second day, and so on, doubling the amount each day. Which arrangement should Samantha choose?

26. Flashing the Signs

Place the given operational signs between the numerals to make true statements.

+	3 9 2 5 1 7 8 = 242
+ and −	3 5 9 1 0 5 3 = 257
×	4 9 2 3 7 0 = 344 610
+, −, and ×	8 0 7 3 1 1 5 = 5834

27. Boxed In

Find integers for the boxes that sum to the numbers in the circles between them.

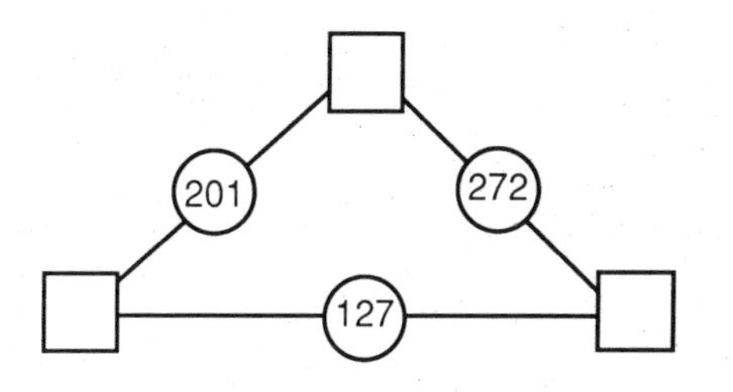

28. Fractionated

What does this picture show?

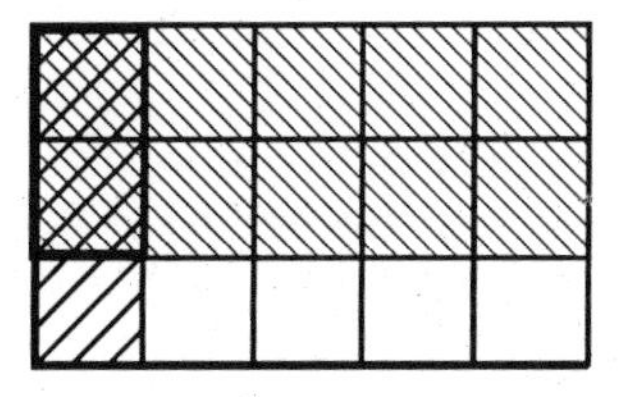

29. Does It Fit?

A house has outside measurements of 46 feet by 54 feet. If a scale drawing of the house is made on graph paper $8\frac{1}{2}$ by 11 inches, what does 1 inch on the graph paper have to equal in whole numbers of feet so that you can obtain the largest accurate drawing entirely on the paper?

30. Rational Numbers Are Dense

Yes, rational numbers are dense. This means that between any two rational numbers there is always another. With that in mind, find the rational number halfway between $\frac{4}{5}$ and $\frac{2}{3}$, and the one halfway between $\frac{1}{2}$ and $\frac{1}{3}$.

31. Take a Number, Please

Pick two numbers, for example, 2 and 4, in that order. Follow this rule: The next number in the series equals (the last number + 1) divided by (the next to last number). Continue this process until you can make up a rule for what is taking place.

32. Percent Message Link

Take the designated percentage of letters in the given words and put them together in order, to form a new word: first 20% of *mouse*; last 80% of *bathe*; first 75% of *mate*; last $66\frac{2}{3}$% of *tic*; and last 10% of *kilometers*.

33. Pattern Blocks Blocked

Complete this pattern:

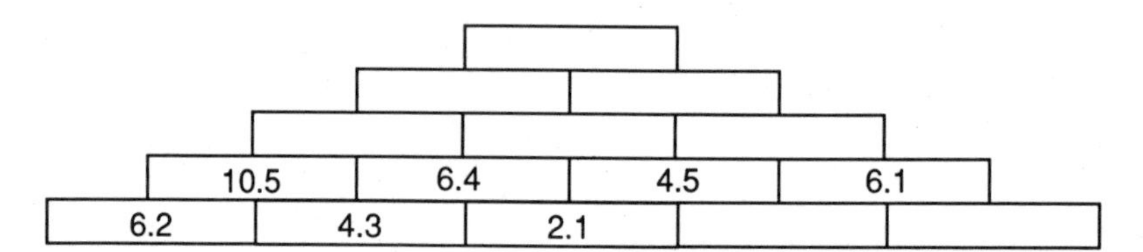

34. Decimated

Place the digits from the left column in the boxes to make the largest number. Do the same to make the smallest.

9, 3, 1, 6	□□.7 + □.□
3, 4, 7, 3	6□.□ − □.□
8, 4, 7, 0	3.4□□ − 2.□□
2, 8, 6, 1	4.□□ × □□

35. Repeating Decimals . . .

When many rational numbers are converted to decimals, they repeat in different ways and lengths. For example, $\frac{1}{3} = .333\ldots$ or $.\overline{3}$. Try these: $\frac{1}{9}$; $\frac{1}{11}$; $\frac{5}{6}$; $\frac{2}{7}$.

36. Converting Repeating Decimals

$.\overline{3}$ is converted to a fraction in this manner:

$$\text{Let } x = .\overline{3}. \qquad \begin{aligned} 10x &= 3.3\ldots \\ -x &= \ \ .3\ldots \\ \hline 9x &= 3. \\ x &= 1/3 \end{aligned}$$

Similarly, find $.\overline{83}$, $.\overline{217}$, $.25\overline{31}$.

37. Population Bulging

A town has a population of 300 000 and an annual growth rate of 4.5 percent. At that rate, in how many years will the population reach 500 000?

"OLDIES BUT GOODIES"

38. Pizza Cutter

Two points on a circle are connected to each other with a segment. How many regions are formed? What number of regions is formed by connecting three points on a circle to each of the other points? Four points? Five points? Six points? (Hint: Make drawings and a chart, and prepare to be surprised.)

39. Farmer's Dilemma

A farmer has a hen, a cat, and a bag of seed to carry across the river. His boat can carry only him and one other item. He cannot leave the cat alone with the hen or the hen alone with the seed. How can he get them all across the river?

40. Logical Disorder

Al, Joe, Carl, and Dave were standing in line. Dave was not first. Joe was between Al and Carl. Al was between Dave and Joe. In what order were they standing in line?

41. $$$$$

In this cryptarithm each letter stands for a certain digit, and each digit has only one letter value. What numerical value does "money" have?

```
  SEND
 +MORE
 -----
 MONEY
```

42. Who Sells What?

Four women sell different perfumes. The women's names are Gail, Pauline, Doris, and Barbara. The perfumes are Passion, Desire, Glob, and Beauty. No one sells a product that begins with the same letter as her first name. Pauline is a friend of the girl who sells Desire. Gail sells Beauty. Who sells which perfume?

43. Single Dip

When Amy, Billy, and Cindy eat out, each one orders either chocolate or vanilla ice cream. If Amy orders chocolate, Billy orders vanilla. Either Amy or Cindy order chocolate, but not both. Billy and Cindy do not both order vanilla. Who could have ordered chocolate yesterday and vanilla today?

44. Let's Be Logical

The following are visual rules:

S ⟶ □	S ⟶̸ △	T*R ⟶̸ ▭
T ⟶ △	$\overset{\frown}{A}$ ⟶ □, △, ▭	T*R ⟶ ▭ △
A ⟶ ○	T*R ⟶ △ ▭	
R ⟶ ▭	T*R ⟶̸ △	

Find the missing element that will make these statements correct.

1) ?*S ⟶ □ ○ 2) $\overset{\frown}{A}$*S ⟶ ?? 3) $\overset{\frown}{A}$*$\overset{\frown}{S}$? ○ □ 4) A*$\overset{\frown}{S}$ ⟶ ??

45. Sports Car Mania

Every red car at an auto show was a sports car. Half of all blue cars were sports cars. Half of all sports cars were red. There were forty blue cars and thirty red cars. How many sports cars were neither red nor blue? (Hint: Try a Venn diagram.)

46. Watch Where You Dig

Three houses are to be connected underground to water, electricity, and gas. For safety reasons, the lines are not to overlap. Can this be done?

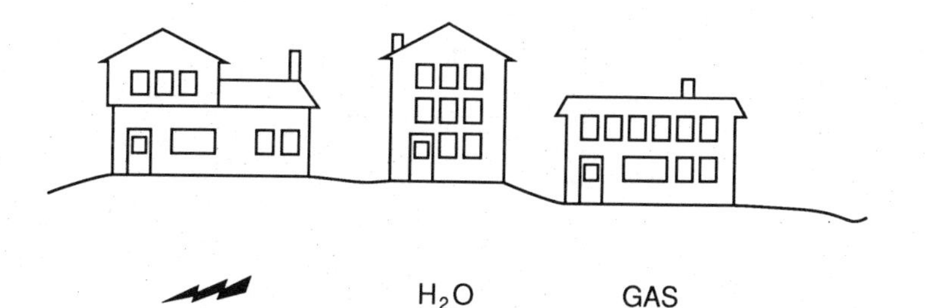

H_2O GAS

47. Bridges to Cross

The town of Königsberg in old Germany was centered on two islands in the middle of a river. Seven bridges connected the various parts of the town to one another. Would it be possible to take a walk across all seven of the bridges without crossing any of them twice?

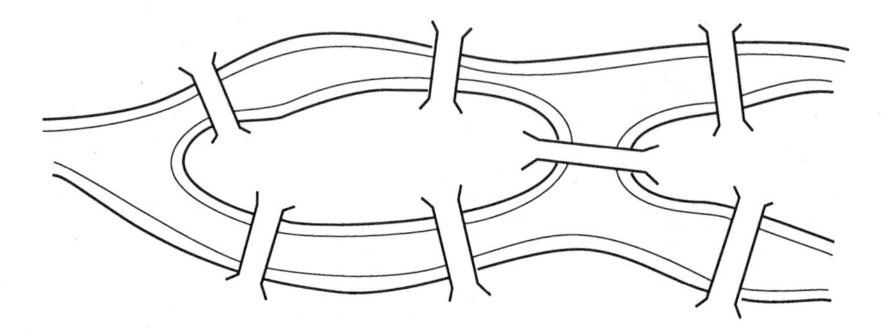

48. Magic Square

Use the remaining numbers from 1 to 16 to complete this magic square in which all rows, columns, and diagonals must sum to the same number.

	2	3	
5	11		8
9		6	
	14		1

49. Magic Squared

This magic square has rows, columns, and diagonals whose four numbers sum to 34. Find other sets of four numbers in this square that also sum to 34. (They form some very interesting patterns.)

16	2	3	13
5	11	10	8
9	7	6	12
4	14	15	1

50. Bucket Brigade

Suppose you have a 3-quart and an 8-quart pail with no markings on either pail. How can you use these pails to get exactly 4 quarts of water in the larger pail?

51. Don't Slam Those Lockers

In a certain junior high school there were 1000 students and 1000 lockers. At the end of each year, the students lined up in alphabetical order and performed the following strange ritual: The first student opened every locker. The second student went to every second locker and closed it. The third student went to every third locker and changed it (i.e., if the locker was open, the student closed it; if it was closed, the student opened it). In a similar manner, the fourth, fifth, sixth, . . . student changed every fourth, fifth, sixth, . . . locker. After all 1000 students had passed by the lockers, which lockers were open?

52. Unbalanced

You have nine identical-looking balls, but one weighs slightly less than the others. Find the light one with the fewest uses of a balance scale.

53. Keeping Your Weight Down

If whole number gram weights are to be put on one pan of a balance and objects to weigh on the other, list the minimum number of weights needed to weigh every whole number of grams from 1 to 63.

54. Koala Klimb

A sleepy koala wants to climb to the top of a eucalyptus tree that is 10 meters tall. Each day the bear climbs up 5 meters, but at night, while asleep, it slides back 4 meters. At this rate, how many days will it take the bear to reach the top of the tree? (Try a picture.)

55. Old McDonald

A farmer has some hens and some rabbits. Together these animals have eight heads and twenty-two feet. How many of each did the farmer have?

56. Coins in a Row

Draw a picture showing how you can arrange six coins in three straight rows of three each.

57. Time after Time

Some time after midnight but before noon of the same day, the hour hand and the minute hand of Ben's watch pointed in the same direction. It happened once in the third hour after midnight. To the nearest second, when did it happen during that hour? How long did Ben wait for the watch's hands to point in the same direction again?

58. Down the Lazy River

A motorboat travels at 25 km an hour in still water. The Cedar River has a current of 5 km an hour. What is the average speed for a trip of 60 km downstream and then back to the starting point?

59. Bike Ride

Jack and Jill live 126 kilometers apart. They want to leave their homes at the same time, ride their bikes toward each other, meet for lunch at 12 o'clock, and then go climb a hill. Jack rides 18 kilometers an hour. Jill rides 24 kilometers an hour. What is the latest time they can leave home?

60. Making Change Count

How many ways can you be given change worth fifty cents?

61. Take a Break

Peggy is writing the numbers from 1 to 1000. She stops to rest after writing 630 digits. What is the last number she wrote?

62. Lay an Egg

Three hens lay four eggs in five days. How many days will it take a dozen hens to lay four dozen eggs?

63. Elevator Ride

Jim got into an elevator. He went down five floors, up six floors, and down seven floors. He was then on the second floor. Where did he get on?

64. Next, Please

Continue this sequence: 1, 2, 5, 26, ?, ?

65. What's the Number?

Find the number that is a three-digit perfect cube and whose digits sum to a perfect square.

66. Guess My Number

I am less than 25. My ones digit is twice my tens digit. My digits add up to an even number. What am I?

SPECIAL COUNTING

67. Caged Mice

How many ways can you put five mice in two cages? Three cages?

68. How Many Whats?

How many s are in ?

69. Squared Away

How many rectangles are in this row of squares?

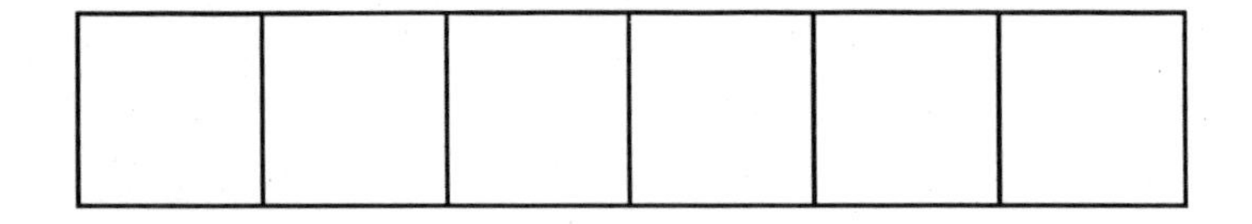

70. Checkerboard Jumps

How many squares are on a 3 × 3 checkerboard? 4 × 4? 8 × 8? (Hint: Start with 1 × 1 and work your way up, looking for patterns.)

71. String Art Unstrung

String art is made by using string to connect evenly spaced nails on a vertical axis with similarly spaced nails on a horizontal axis, as in the examples. How many intersections are made by the strings connecting eight nails on each axis? *n* nails? (Hint: Start with a smaller number of nails and a table for your results.)

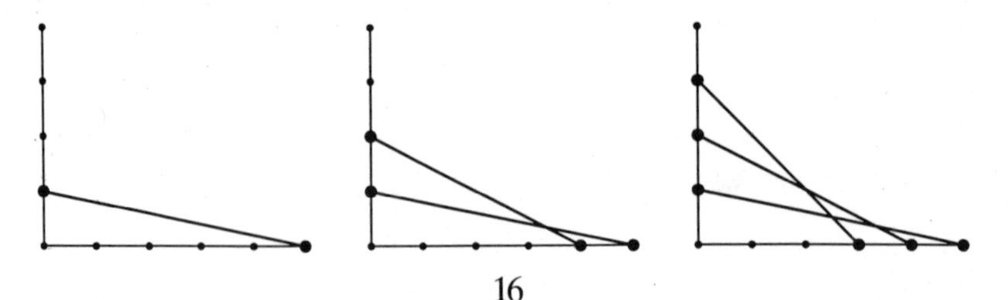

72. Falling Dominoes

How many dominoes are in a complete set when six is the maximum number of dots on a domino half?

73. Moving from *A* to *B*

Imagine a girl at point *A* who wants to see her friend at point *B*. If the girl can only move lower on the diamond grid along the streets shown, how many ways can she go? (Hint: Think of Pascal's triangle.)

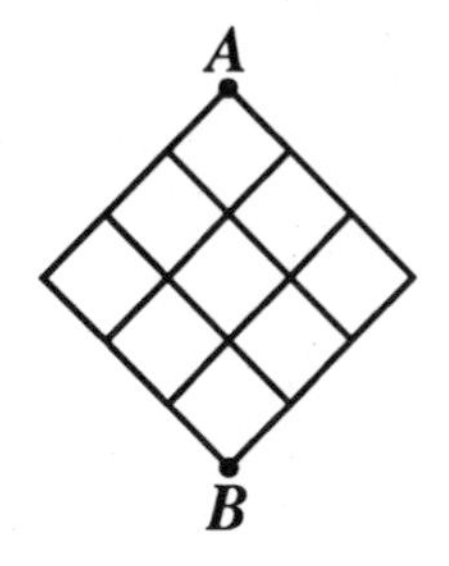

74. Array, Array

How many different ways can you place four X's in a 4×4 square grid so that each row and column has only one X in it? How many of these solutions have no more than one X in any diagonal?

<table>
<tr><td></td><td>X</td><td></td><td></td></tr>
<tr><td></td><td></td><td></td><td>X</td></tr>
<tr><td>X</td><td></td><td></td><td></td></tr>
<tr><td></td><td></td><td>X</td><td></td></tr>
</table>

75. Staircase Fall

Ten blocks are needed to make a staircase of four steps. How many blocks are needed to make ten steps? How many blocks are needed to make fifty steps?

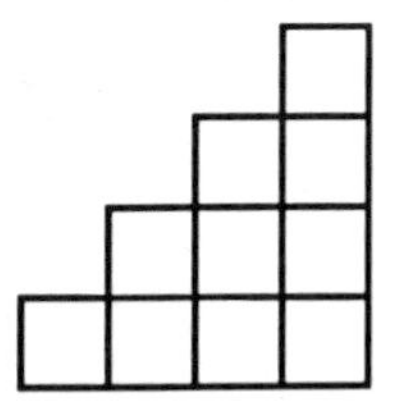

76. High-Rise Building

Here are three buildings in a series made of cubes. How many cubes are in the twentieth building?

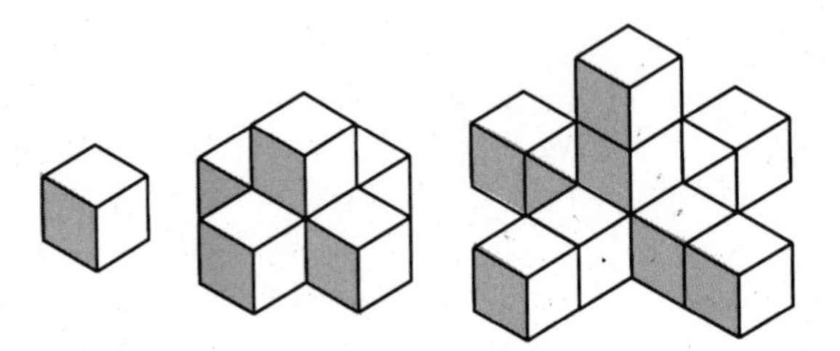

77. Dot to Dot

In which of these two drawings can each point be connected to other points so that each point has exactly three connections?

OFFBEAT AND UNUSUAL

78. Pictographic

What does this pictogram show?

1,000, **1** 000

79. Mystery Letters

What are the next three letters in this sequence? O, T, T, F, F, S, S, ?, ?, ?

80. What Comes Next?

Complete this sequence: 11, 31, 71, 91, 32, 92, 13, 73, 14, 34, 74, 35, 95, 16, 76, 17, 37, 97, 38, 98, ?

81. Calculator Words

On a calculator the following numbers can be read as letters when the calculator is turned upside down.

8→B 4→h 5→S 3→E 1→I 6→g 0→O

These letters can be used to form words that are answers to "numerical" questions such as the following:

a) What do you call big pigs?
$(500 \times 10) + (20 \times 30) + (2^2)$

b) You're a good __________.
$(2 \times 3) \times 100 + (7 \times 9)$

82. Mathematical "Equations"

Each "equation" is in the form of a number phrase with the key words abbreviated. Solve the following:

365 = D. in a Y.
360 = D. in a C.
4 = S. in a Q.
3.14 = P. as a D. (R.O.)

83. Believe It or Not

Is it possible to subtract 1 from 19 and get 20? Can ½ of 13 = 8?

84. Valentine Day Message

Decode this message: XPVS NBUI UFBDIFS MPWFT ZPV.

85. Digital "Burn Out"

The face of a 12-hour digital clock can be considered to be constructed of twenty-three different lights and a colon. What time is it when the fewest number of lights are lit? What time is it when the largest number of lights are lit?

18:88

86. Timely Problem

In 24 hours on a digital clock, how many times does at least one nine show?

87. Insomnia

On a digital clock readout, how many times between midnight and 7:00 a.m. contain consecutive digits? How many times after 7:00 a.m.?

GEOMETRY

88. Which Solid Am I?

Identify the following solids:

a) I have six flat faces. I have twelve edges. I have eight vertices.
b) I have five flat faces. I have nine edges. I have six vertices.
c) I have two flat faces. I can roll.
d) I have five flat faces. I have eight edges. I have five vertices.

89. In the Middle of Things

For each of these five figures, find the midpoint of each side. Then connect the midpoints of adjacent sides. What pattern do you notice?

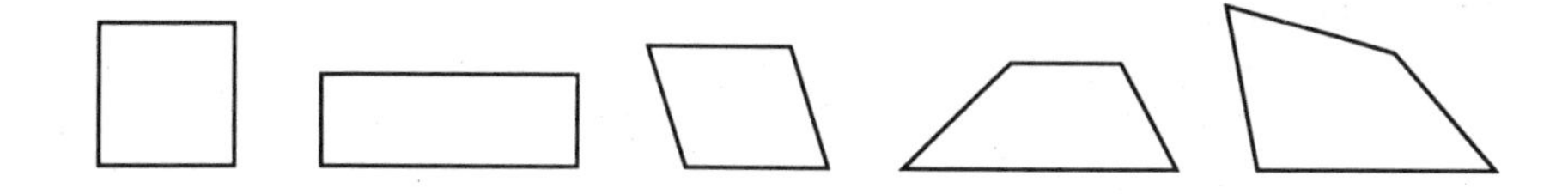

90. Polyhedra Patterns

Fill in this chart and generalize the results:

	Vertices (V)	Edges (E)	Faces (F)	$V - E + F$
Cube				
Pentagonal pyramid				
Triangular prism				

91. Making Clones

Divide the parallelogram into eighths and the triangle into fourths.

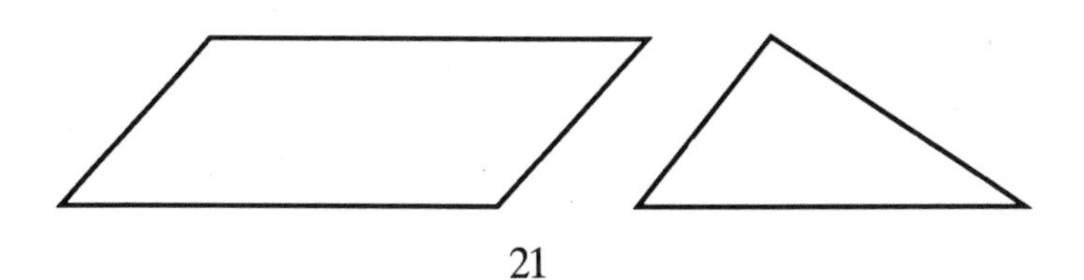

92. Cross-section Slice

A cube is sliced by a saw. What kinds of cross sections can result?

93. Dice-o-matic

A $3 \times 3 \times 3$ rectangular solid is painted and then sliced into cubes one unit on an edge. How many of the resulting cubes have no faces painted? One face painted? Two faces painted? Three faces painted? Four or more painted faces?

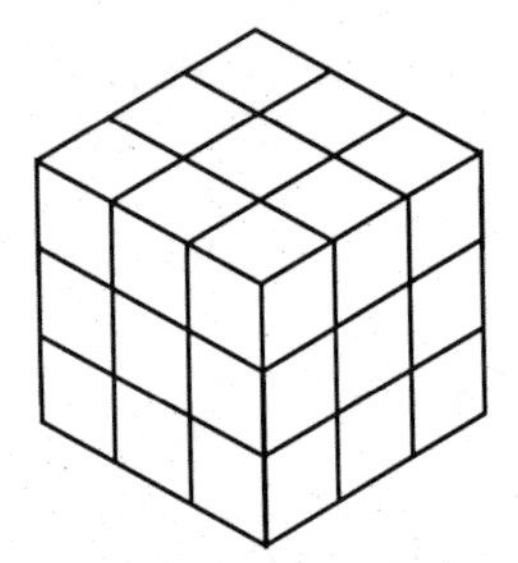

94. Triangulate

List all the triangles in this figure.

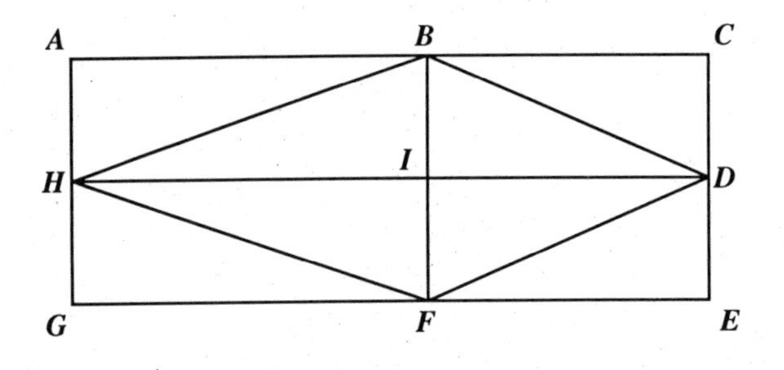

95. It Was Here Before

Pirate Pete remembered burying his treasure on Buccaneer Island 60 paces from the tall palm and 75 paces from the large rock. When he returned, measured off distances carefully, and dug, there was no treasure. What went wrong?

96. A Square Deal

(6,3) and (3,6) are vertices of a square. Find the other two. Consider all the possibilities.

97. Baseball with Square Objects

A baseball diamond is a square with a perimeter of 360 feet. The distance from the pitcher's mound to home plate is 6 inches more than ⅔ of the distance between consecutive bases. When the pitcher's foot is on the mound, how many feet is the pitcher from home plate?

98. Round and Round We Go

A bike has wheels with a 10-inch radius. How many revolutions do the wheels make in a mile ride? (Hint: Use $\pi = \frac{22}{7}$.)

99. Speedy Earth

Assuming the earth is 93 000 000 miles from the sun and has an almost circular orbit, what is its speed in orbit?

100. Find My Area

Find the area of this figure. Try to do it in more than one way. Assume all angles shown are right angles.

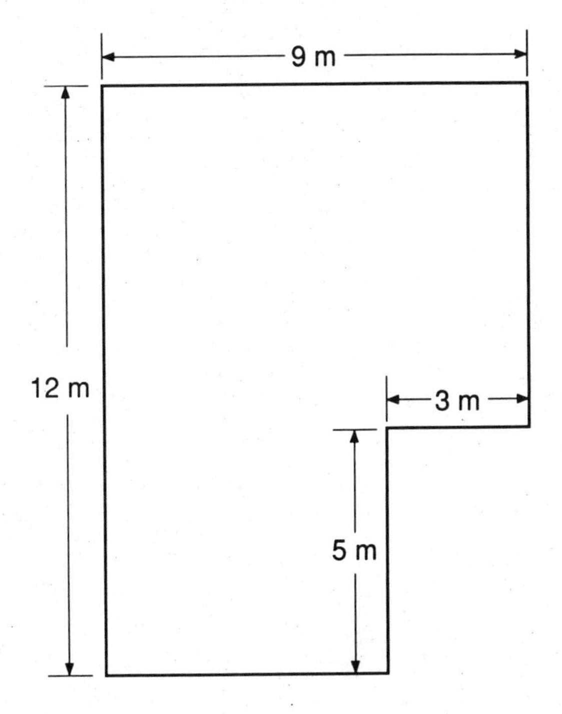

101. Rectify This Rectangle

The 8×8 square is cut apart and reassembled as this 5×13 rectangle. How can you justify this?

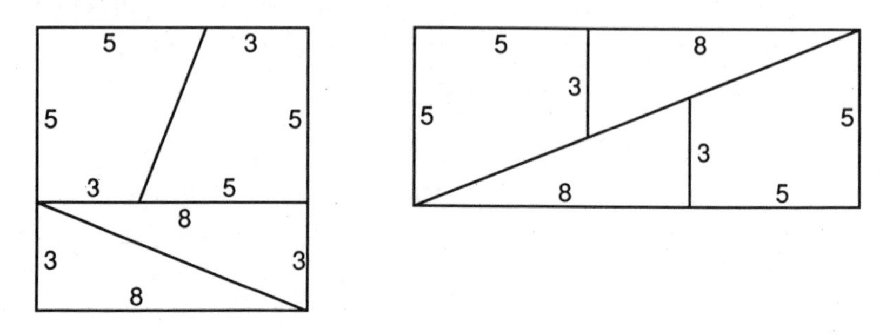

102. Stay off the Grass

Find the area of this park without the sidewalk.

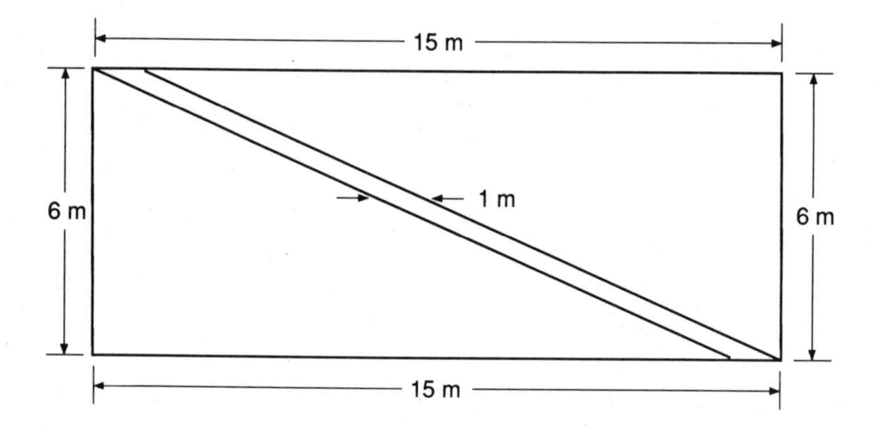

103. Pi Are Square

If the radius of the circle inscribed in the square is 2.2 centimeters, what is the area of the shaded region? (Use $\pi = 3.14$.)

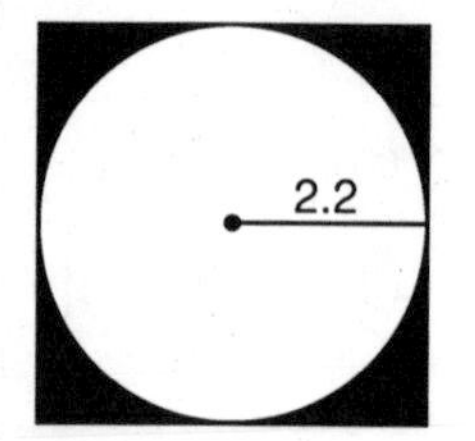

104. Crazy as a Lune

AOB is a quadrant of circle *O*. Chord *AB* is the diameter of semicircle *ACB*. The area bounded by the arc *AB* and the semicircle ACB is called a lune. Compare the area of triangle *AOB* and the area of the lune.

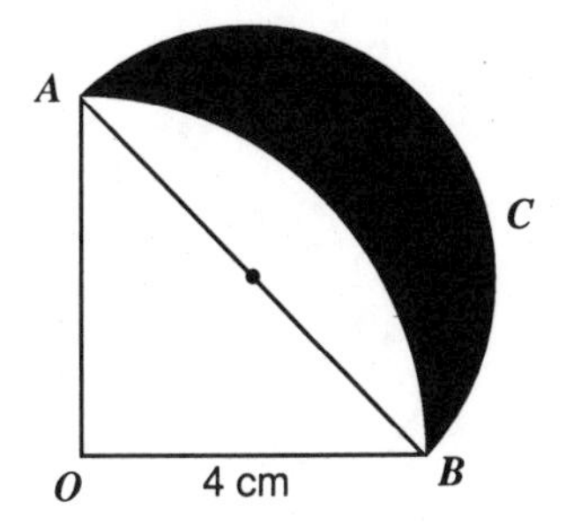

105. Graph-Paper Polygons

Draw on graph paper (or construct on a geoboard) the following figures: (1) a rectangle with an area of 6 square units; (2) an octagon with an area of 10 square units; and (3) a polygon with an area of 8 square units, but the least perimeter.

106. Geoboard Areas

Find the area in square units of each of these three shapes:

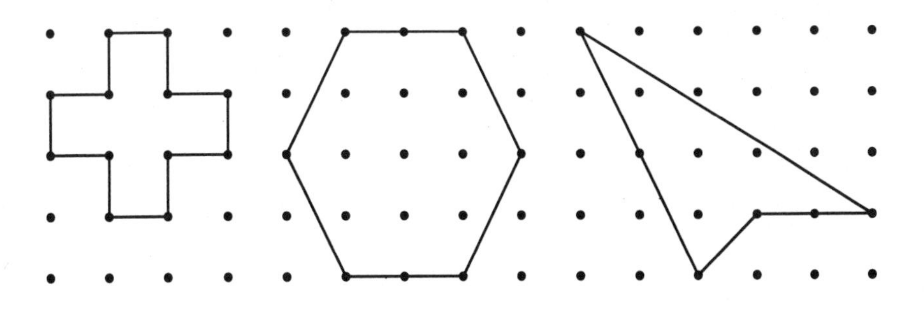

107. Pick Your Way through This

Using dot paper, determine a formula showing the relationship between the area (A) of any closed polygon, the number of dots on the polygon (x), and the number of dots inside the polygon (y). This relationship is known as Pick's formula.

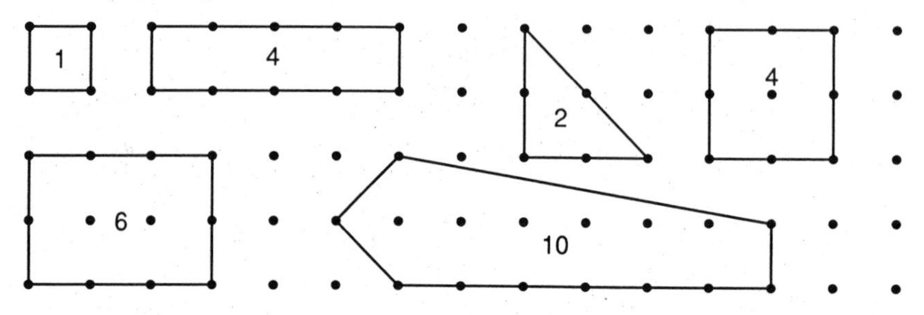

108. Mirror My Thoughts

Use a ruler and protractor to make this design symmetric.

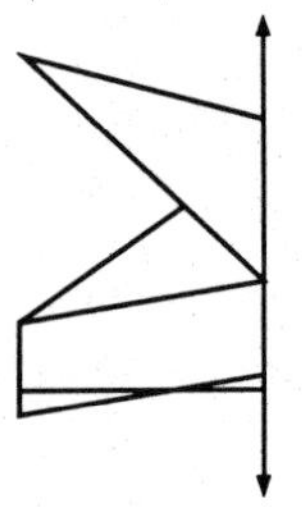

109. A Capital Idea

Which uppercase letters of the alphabet have line symmetry, that is, have at least one place where you can fold half of the letter exactly on top of itself?

110. 3 × 3 Symmetry

Using 0 to 4 colored squares in this 3 × 3 grid, how many distinct patterns have exactly one line of symmetry?

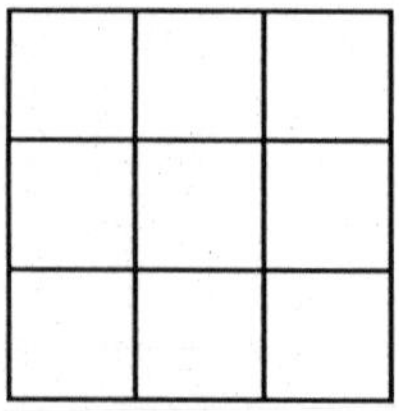

111. Regular Polygon Connection

Fifty squares, each with edge 1 unit long, are placed side by side in a row. What is the perimeter of the resulting figure? What if the squares are replaced with equilateral triangles with edges of 1 unit? Regular hexagons with edges of 1 unit?

112. Building Blocks

You have twelve identical cubes. How many rectangular solids can be formed that use all twelve? Which of these has the least surface area? Which has the greatest surface area?

113. Building a Bigger Box

A piece of paper measures 12 centimeters on a side. Imagine cutting an equal-sized square out of each corner and folding to make an open box. What size squares should you cut to make the box with the largest volume?

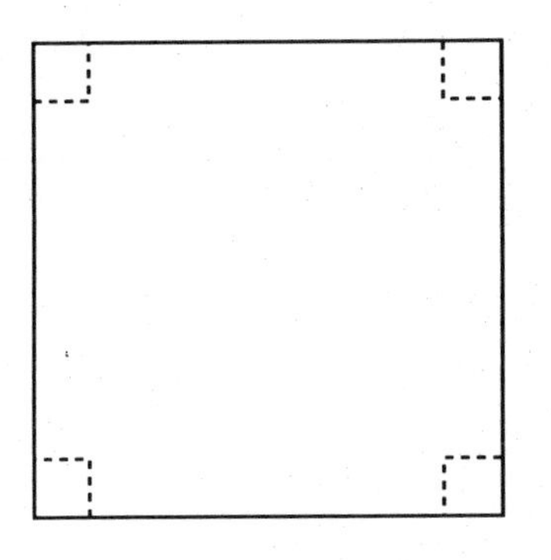

114. The Pentomino Theory

A pentomino is a plane figure that is formed by grouping five congruent squares so that every square has at least one of its sides in common with at least one other square. Below are two examples. How many more are there?

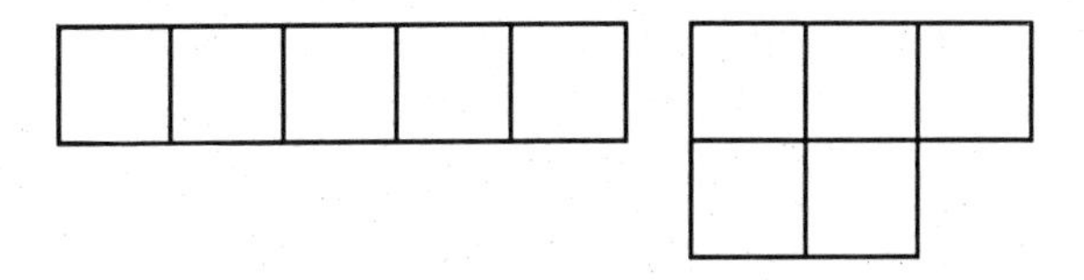

115. In Black and White

A square is divided into eight equal parts as shown. If repeated patterns due to flips or spins are not counted, how many different ways can you color half of the triangular regions?

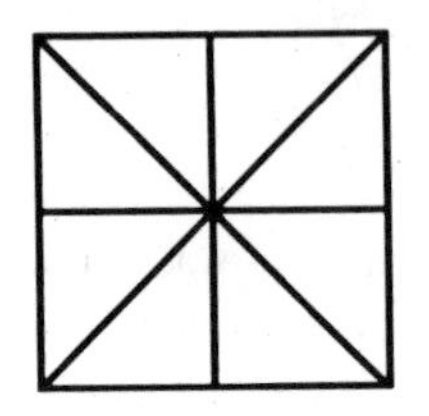

116. Encircled

Find all possible quadrilaterals that can be made by connecting eight equally spaced points on a circle.

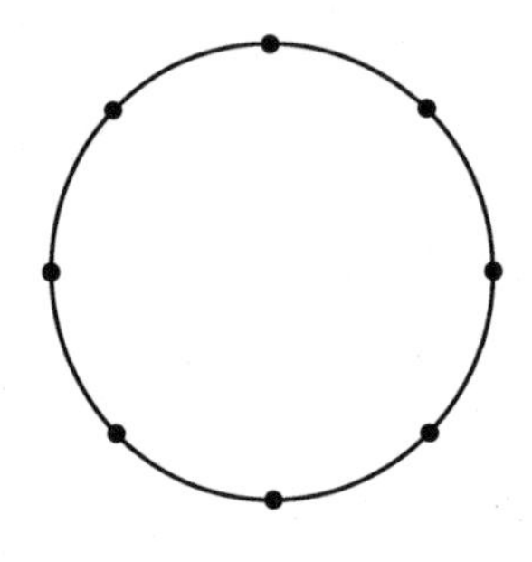

117. "Circular" Parallelograms

The circle below was cut into parts and reassembled as a "parallelogram." In terms of the circle, what is the parallelogram's height? Base? Area?

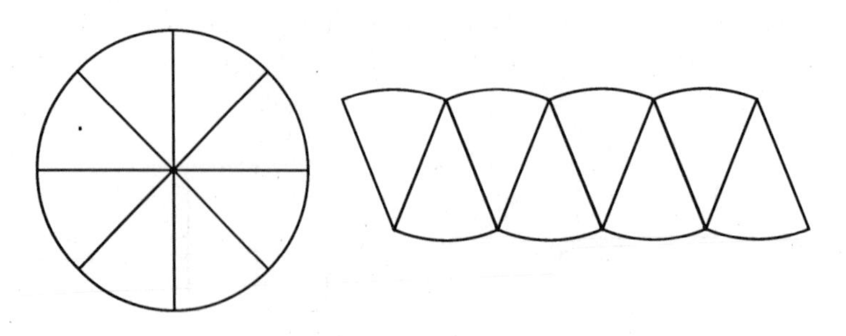

118. Sharing a Doughnut

How can you cut a doughnut into 8 equal pieces with three cuts of a knife? Into 12 unequal pieces with three cuts?

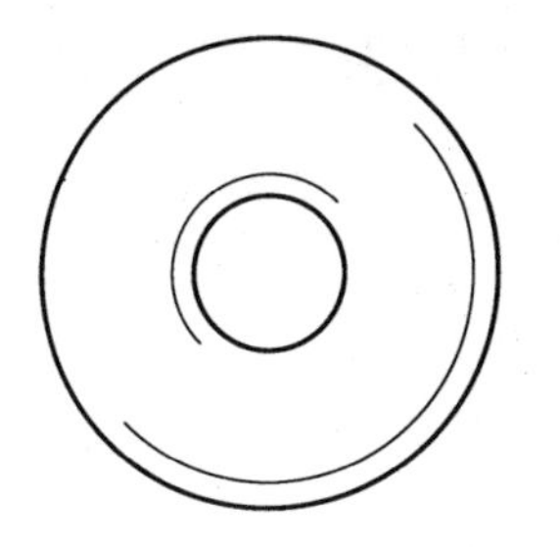

PROBABILITY

119. Tennis, Anyone?

In a tennis tournament 66 matches were played. If each person played each other member once, how many people played in the tournament?

120. Share a Scare

Eric and six friends spent a day at an amusement park. At the end of the day they decided to pair up for roller-coaster rides so that each friend would ride with each of the others exactly once. How many rides must be taken?

121. Roll Those Dice

Roll a pair of regular dice. What is the lowest total you can get? What is the highest? What number is most likely to appear? What fraction of the time does this most likely event occur?

122. A Fishy Tasting Sample

Twenty-four tagged fish are added to a pond containing fish. Later a sample of ten fish is taken, and two are found to be tagged. Estimate the fish population of the pond.

123. Book Odds

A book has 167 pages with page 1 on the right side of the book and page 2 on the left. What is the probability that when you open the book, one of the page numbers is a multiple of 3? Or that on exactly one page the sum of the digits is greater than 10?

124. Repeating 3s

With the digits 1, 2, 3, 4, and 5, how many five-digit positive integers can be formed if only the 3 can be repeated any number of times?

125. Triangle à la Pascal

Continue this pattern for the next two rows. This is known as Pascal's triangle in honor of the great French mathematician. It has many interesting properties.

1

1 1

1 2 1

1 3 3 1

1 4 6 4 1

1 5 10 10 5 1

ANSWERS/REFERENCES

1. The result is always 1089. As examples: $432 - 234 = 198$ and $198 + 891 = 1089$; $100 - 001 = 099$ and $099 + 990 = 1089$.

(Ewbank and Ginther, January 1984, pp. 49–51; Crawford, January 1985, p. 4; Flexer, September 1979, pp. 22–26)

2. $496 = 1 + 2 + 4 + 8 + 16 + 31 + 62 + 124 + 248$.

(Hopkins, March 1987, pp. 38–42; O'Daffer, October 1981, p. 27)

3. 2; odd number > 1.

(Swafford and McGinty, October 1978, pp. 16–17; Scott, December 1981, p. 4; Hohlfield, December 1981; pp. 28–29)

4. Factors of 220 are 1, 2, 4, 5, 10, 11, 20, 22, 44, 55, and 110, which sum to 284. Factors of 284 are 1, 2, 4, 71, and 142, which sum to 220.

(Hopkins, March 1987, pp. 38–42; Hiatt, February 1987, pp. 38–43)

5. 13, 21, 34.

(Payne, May 1987, p. 31; Beattie, April 1979, pp. 36–38)

6. (a) $32 + 23 = 55$; (b) $458 + 854 = 1312$, $1312 + 2131 = 3443$; (c) $748 + 847 = 1595$, $1595 + 5951 = 7546$, $7546 + 6457 = 14\,003$, $14\,003 + 30\,041 = 44\,044$.

(Whitin, November 1985, pp. 25–26; Schaaf, March 1986, p. 35; Dockweiler, January 1985, p. 46; Nichol, December 1978, pp. 20–21; Robinson, September 1979, p. 54; Trotter, November 1979, p. 52)

7. 9 takes nineteen steps.

(Rubillo, November 1987, pp. 54–55)

8. $n(n + 1)/2$.

(Speiler, November 1984, p. 44; Anderson, November 1983, pp. 50–51)

9. $1 + 6 = 7$, $7 + 12 = 19$, $19 + 18 = 37$, $37 + 24 = 61$, $61 + 30 = \mathit{91}$, $91 + 36 = \mathit{127}$.

(Standen, January 1988, p. 40; Spikell, September 1988, p. 5)

10.

592

444 370 222 740

148 74

296

666 888

(Vangrin, November 1976, pp. 492, 577; Locke, May 1977, pp. 416–17; Flexer, September 1979, pp. 22–26)

11.

19
1
8 9
6 2
4 5 3 7

20
4
8 1
3 9
5 7 2 6

21
3
8 7
4 2
6 5 1 9

23
7
3 6
5 1
8 2 4 9

(Yates, May 1976, pp. 351–54; Rudnytsky, November 1976, pp. 524–25; Bohland, November 1976, pp. 525–26; Roessel, December 1977, p. 22; Flexer, September 1979, pp. 22–26)

12.

3
19 17
16 7 18
2 1
12 5 11
4 6
10 8 9
13 14
15

(Krulik, April 1980, pp. 40–42)

13. 364 = 1 + 3 + 6 + 10 + 15 + 21 + 28 + 36 + 45 + 55 + 66 + 78.

(Stern, December 1983, p. 3)

14.

	18	—	55
—	—	73	101
—	97	106	—
33	51	—	88

	—	44	—
—	1909	3652	7055
56	—	—	4760
99	2277	4356	—

(Erb, September 1988, p. 39)

15. Six steps. This occurs when the hundreds digit minus the units digit equals 2 or 9.

(Padberg, April 1981, pp. 21–23)

16. Diagonal 1 has counting numbers. The last term in each row n is n. In diagonal n you count by n. Many others are possible.

(Ouelette and Gannon, January 1979, pp. 34–38)

17.

1	32√
2	64√
4	128
8	256
16	512√
	608

This method works by taking one factor and successively doubling it. The numbers in the first column required to sum to the other factor are noted, and the corresponding doubles are added.

(Arcavi, December 1987, pp. 13–16)

18. Start with the 11 on the bottom row, go to 17, go to 7, go to 3, and exit.

(Lappan and Winter, March 1980, pp. 24–27)

19. The difference is always 8. For example, using 10, 12, 14, and 16; $12 \times 14 = 168$, $10 \times 16 = 160$, and $168 - 160 = 8$.

(Hastings, May 1987, p. 61)

20. The product of the tens digit must equal the product of the ones digits.

(Schaaf, January 1986, pp. 38–39)

21. $642 \times 83 = 53\ 286$; $24 \times 368 = 8832$.

(Barson and Barson, October 1987, pp. 28, 31; Reisberg et al., January 1988, p. 3; Schaaf, March 1986, pp. 34–35)

22. 126, 127.

(Wilmot and Barnes, March 1988, pp. 32–33; Ockenga and Duea, March 1979, pp. 28, 30)

23. In September 1988, "typey" and "teaette" were each reported to sum to 1 000 000.

(Bain, September 1987, p. 26; Neidenbach, May 1988, p. 4; Lucarelli, April 1978, p. 43)

24. 15. 5/18/90, 6/15/90, 9/10/90, 10/9/90, 7/13/91, 4/23/92, 3/31,93, 5/19/95, 4/24/96, 6/16/96, 8/12/96, 12/8/96, 7/14/98, 9/11/99, and 11/9/99.

(Trotter, October 1983, p. 53)

25. The doubling alternative earns $55.35 more.

(O'Daffer, January 1985, p. 34)

26. $39 + 25 + 178 = 242$; $359 - 105 + 3 = 257$; $4923 \times 70 = 344\ 610$; $80 \times 73 - 11 + 5 = 5834$.

(Shaw, April 1985, pp. 28, 31)

27. 173, 28, 99.

(Ewbank and Ginther, January 1984, pp. 49–51)

28. $\frac{2}{3} \times \frac{1}{5} = \frac{2}{15}$.

(Edge, April 1987, pp. 13–17; Prevost, February 1984, pp. 43–46; Thornton, February 1977, pp. 154–57; Ockenga and Duea, January 1978, pp. 28, 32)

29. Six factors.

(Coppel, February 1977, pp. 125–26)

30. $\frac{11}{15}$; $\frac{5}{12}$.

(Skypek, February 1984, pp. 10–12; Schroeder, November 1983, pp. 3–4; Modell, December 1982, pp. 50, cover 3; Spragens, October 1983, p. 6)

31. The pattern repeats in blocks of five. In our example we get <u>2, 4, 5/2, 7/8, 3/4,</u> (2, 4, . . .)

(Snover, March 1982, pp. 22–26)

32. "Mathematics."

(Leiva, February 1985, p. 3)

33.

49.3
27.8 21.5
16.9 10.9 10.6
10.5 6.4 4.5 6.1
6.2 4.3 2.1 2.4 3.7

The block above is the sum of the two immediately below.

(Williams, September 1984, pp. 40–43; Tabler and Jacobson, December 1980; pp. 19–24)

34. 96.7 + 3.1 or 93.7 + 6.1; 16.7 + 3.9 or 13.7 + 6.9
67.4 − 3.3; 63.3 − 7.4
3.487 − 2.04; 3.404 − 2.87
4.61 × 82; 4.68 × 12

(Payne and Towsley, March 1987, p. 31)

35. $.\overline{1}$, $.\overline{09}$, $.8\overline{3}$, $\overline{.285714}$.

(Ryoti, February 1987, pp. 62–63; Hobbs and Burris, April 1978, pp. 18–20)

36. 83/99; 217/999; 1253/4950.

(Woodcock, November 1984, pp. 4, 44; Anderson, September 1985, p. 6; Watson, March 1985, p. 44; Robidoux and Montefusco, January 1977, pp. 81–82)

37. Between 11 and 12 years.

(Comstock and Demana, February 1987, pp. 48–51; Mason, April 1986, p. 42)

38. 2; 4; 8; 16; 30 or 31.

(Thompson, September 1985, pp. 21–23)

39. Take the hen across. Return. Take the cat across. Return with the hen. Take the seed across. Return. Take the hen across.

(Wilmot and Barnes, March 1988, p. 33; Barson and Barson, November 1987, pp. 28, 32)

40. Carl, Joe, Al, Dave.

(Barson and Barson, November 1987, p. 32)

41. 10 652.

(Edge, November 1986, pp. 12–15)

42. Gail-Beauty, Pauline-Glob, Doris-Passion, Barbara-Desire.

(Barson and Barson, November 1987, pp. 28, 32; O'Daffer, February 1985, p. 62; Masse, December 1978, pp. 11–14)

43.

		yesterday	*today*
Billy, but only when:	Amy	vanilla	vanilla
	Billy	chocolate	vanilla
	Cindy	chocolate	chocolate

(Warman, May 1982, pp. 26–30)

44. 1) A; 2) □□ or △□ or ▭□; 3) ⟶; 4) ○○ or ○△ or ○▭.

(Murphey, March 1983, pp. 28, 49)

45. 10.

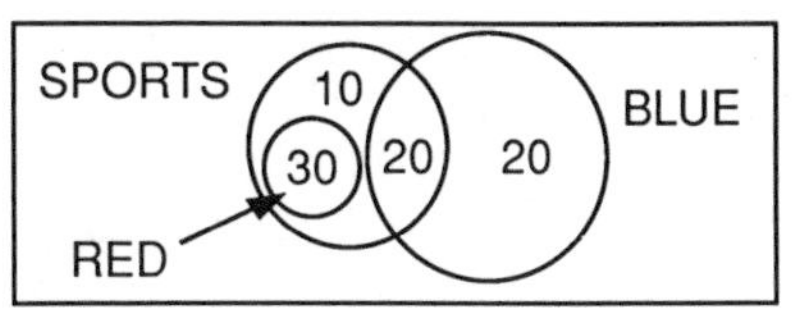

(O'Daffer, February 1985, p. 63)

46. No.

(Avital and Grinblat, March 1983, pp. 46–49)

47. No.

(Slesnick, March 1984, pp. 41–43; Zaslavsky, October 1981, pp. 42–47)

48.

16	2	3	13
5	11	10	8
9	7	6	12
4	14	15	1

(Barson and Barson, February 1988, p. 35; Sherill, October 1987, pp. 44–47; Sullivan, October 1976, pp. 427–28; Lott, March 1977, pp. 228–29; Bright, October 1977; pp. 30–34; Ginther, October 1978, p. 52; Pagni, May 1974, pp. 439–41)

49.

16	–	–	13
–	–	–	–
–	–	–	–
4	–	–	1

and many others.

(Hiatt, February 1987, p. 39)

50. Fill 8-quart pail. Fill 3-quart pail from 8. Dump 3. Fill 3 from 8. Dump 3. Pour 8 into 3. Refill 8. Fill 3 from 8. Dump 3. Fill 3 from 8.

(O'Daffer, April 1985, p. 35; Lundberg, September 1985, p. 6; Jensen and O'Neil, January 1982, pp. 8–9)

51. Perfect squares: 1, 4, 9, 16, 25, . . ., 900, 961, or one open, two closed, one open, four closed, one open, six closed, . . ., one open, sixty closed, one open.

(House, October 1980, pp. 20–23)

52. Two weighings are enough. Put any three balls on each pan of the balance. Case 1: If they are in balance, one of the other three is light. Put one of the remaining ones on each pan of the balance. If they do not balance, you've found the light one; if they do balance, the last one is the light one. Case 2: If the first six balls are not in balance, take the lighter three. Put one of the three on each pan and proceed as above.

(Willcut, February 1980, pp. 16–19)

53. Six weights. 1, 2, 4, 8, 16, 32.

(Adkins, November 1980, pp. 48–49)

54. Six days.

(O'Daffer, October 1984, p. 42; Moses, December 1982, p. 14; Jensen and O'Neil, January 1982, pp. 8–9)

55. Three rabbits, five hens.

(Lee, January 1982, p. 17)

56.

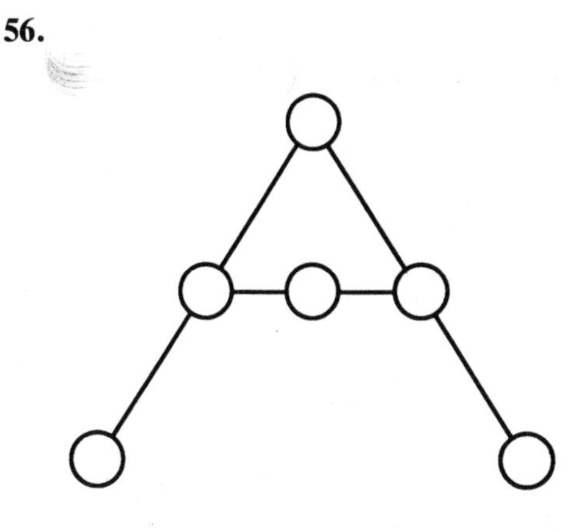

(Payne, May 1987, p. 32)

57. The minute hand travels 6° per minute whereas, the hour hand travels 30° per hour, or half a degree per minute. Between 2 o'clock and 3 o'clock the hour hand has a 60 degree headstart. Setting the degrees equal at t minutes, we get $6t = 60 + t/2$, or $t = 10\ 10/11$ minutes. The first time it happens is 2:10.55 a.m. The next time the hands coincide is 3:16.22 a.m., or 1 hour, 5 minutes, and 27 seconds later.

(Murphey, April 1983, pp. 24, 33)

58. 24 kilometers an hour.

	Down	*Up*	*Total*
d	60 km	60 km	120 km
r	25 + 5 km/hr	25 − 5 km/hr	? km
t	2 hr	3 hr	5 hr

(Leutzinger, October 1985, pp. 34–35)

59. 9:00 a.m.

$$d = \text{rate of Jack} \times \text{time} + \text{rate of Jill} \times \text{time}$$
$$126 = 18t + 24t$$
$$126 = 42t$$
$$t = 3 \text{ hours}$$

(O'Daffer, May 1985, p. 15)

60. Fifty.

$ .50—	1	0	0	0	0	0	0	0	0	0	0	0	0	0	0
.25—	0	2	1	1	1	1	1	1	1	1	1	1	1	1	0
.10—	0	0	2	2	1	1	1	1	0	0	0	0	0	0	5
.05—	0	0	1	0	3	2	1	0	5	4	3	2	1	0	0
.01—	0	0	0	5	0	5	10	15	0	5	10	15	20	25	0

0	0	0	0	0	0	0	0	0	0	0	0	0	0	0
0	0	0	0	0	0	0	0	0	0	0	0	0	0	0
4	4	4	3	3	3	3	3	2	2	2	2	2	2	2
2	1	0	4	3	2	1	0	6	5	4	3	2	1	0
0	5	10	0	5	10	15	20	0	5	10	15	20	25	30

0	0	0	0	0	0	0	0	0	0	0	0	0	0	0
0	0	0	0	0	0	0	0	0	0	0	0	0	0	0
1	1	1	1	1	1	1	1	1	0	0	0	0	0	0
8	7	6	5	4	3	2	1	0	10	9	8	7	6	5
0	5	10	15	20	25	30	35	40	0	5	10	15	20	25

0	0	0	0	0
0	0	0	0	0
0	0	0	0	0
4	3	2	1	0
30	35	40	45	50

(Burns, May 1988, pp. 30–31)

61. 246.

(Leutzinger, September 1985, p. 39)

62. A dozen hens would lay sixteen eggs in five days and forty-eight eggs in fifteen days.

(Payne, May 1988, p. 32)

63. Eighth floor.

(Duea and Ockenga, September 1986, pp. 34–35)

64. 677, 458 330. Square the number and add 1.

(Schueler and Williams, March 1980, p. 28)

65. $216 = 6 \times 6 \times 6$ and $2 + 1 + 6 = 9 = 3 \times 3$.

(Greenes and Shulman, October 1982, pp. 18–21)

66. 24.

(Barson and Barson, November 1987, p. 32)

67. Two cages allow 6 ways.

A—5 4 3 2 1 0
B—0 1 2 3 4 5

Three cages allow 21 ways.

A—5 4 4 3 3 3 2 2 2 2 1 1 1 1 1 0 0 0 0 0 0
B—0 1 0 2 1 0 3 2 1 0 4 3 2 1 0 5 4 3 2 1 0
C—0 0 1 0 1 2 0 1 2 3 0 1 2 3 4 0 1 2 3 4 5

(Ockenga and Duea, April 1977, pp. 303–4)

68. 12.

(Dana and Lindquist, September 1978, pp. 2–9)

69. 21.

number of squares in a row	1	2	3	4	5	6
number of squares	1	2	3	4	5	6
number of nonsquare rectangles	0	1	3	6	10	15
total	1	3	6	10	15	21

(Bright, May 1978, pp. 39–43)

70. (1×1) 1; (2×2), $1 + 4 = 5$; (3×3), $1 + 4 + 9 = 14$; (4×4), $1 + 4 + 9 + 16 = 30$; (5×5), $1 + 4 + 9 + 16 + 25 = 55$; (6×6), $1 + 4 + 9 + 16 + 25 + 36 = 91$; (7×7), $1 + 4 + 9 + 16 + 25 + 36 + 49 = 140$; (8×8), $1 + 4 + 9 + 16 + 25 + 36 + 49 + 64 = 204$.

(Day, October 1986, pp. 14–17; Bright, May 1978, pp. 39–43; Whitin, December 1979, pp. 38–39; Maletsky, February 1982, pp. 20–24)

71. 28; $\frac{n(n-1)}{2}$.

(Krulik and Rudnick, December 1985, pp. 38–39)

72. 28 with 7 doubles and $(7 \times 6)/2 = 21$ different pairs.

(Maletsky, February 1982, pp. 21–22)

73.

```
      –
    1   1
  1   2   1
1   3   3   1
  4   6   4
   10  10
     20
```

(Jamski, October 1985, p. 6)

74. 24; 4.

(Murphey, March 1983, pp. 33, 35; Murphy and Scheding, November 1983, p. 6)

75. $1 + 2 + 3 + \ldots + 10 = 55$; $1 + 2 + 3 + \ldots + 50 = 1275$. $n(n + 1)/2$ yields the same results.

(Thompson, September 1985, pp. 20–21)

76. 96 cubes.

(Maier and Nelson, October 1986, pp. 34–35)

77. The first one.

(Avital, September 1986, pp. 42–45)

78. "One in a Million."

(Koontz and Sovchik, November 1987, p. 33)

79. E[ight], N[ine], T[en].

(Payne, May 1987, p. 31)

80. 79. These are reversed two-digit primes.

(*Problematical Recreations,* December 1982, pp. 26, 36)

81. a) hogs; b) egg.

(Lindberg, February 1987, p. 5; Friesen, December 1976, p. 660)

82. Days in a Year; Degrees in a Circle; Sides in a Quadrilateral; Pi as a Decimal (Rounded Off).

(Jamski, January 1986, p. 43)

83. Yes. XIX−I=XX Yes. XIIII → VIII

(Parillo, September 1979, p. 45)

84. Replace each letter with the previous one in the alphabet and you get: "Your math teacher loves you."
(Mort, February 1983, p. 4)

85. 1:11; 10:08.
(Makurat, February 1985, p. 6)

86. 252.
(Greenes and Immerzeel, March 1987, pp. 34–35)

87. 9—12:34, 1:23, 2:34, 3:21, 3:45, 4:32, 4:56, 5:43, 6:54. 0.
(Burruss, May 1988, p. 39)

88. a) cube; b) triangular prism; c) cylinder; d) square pyramid.
(Edge, November 1986, pp. 12–15)

89. Each figure drawn by joining the midpoints is a parallelogram.
(Bright, December 1977, pp. 28, 32)

90.

8	12	6	2
6	10	6	2
6	9	5	2

$V - E + F = 2$ in all cases

(Clements and Battista, February 1986, pp. 29–32)

91.

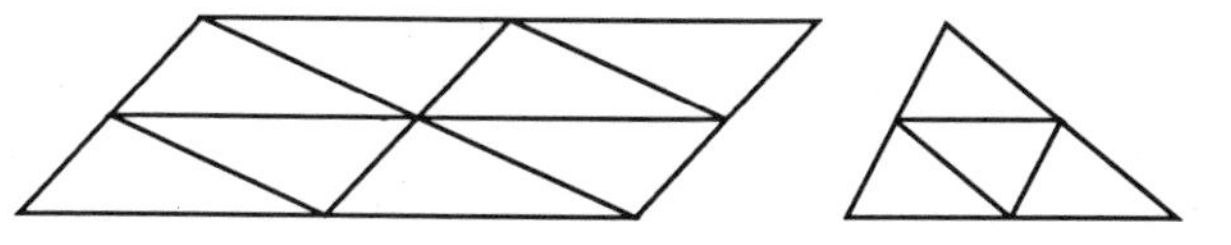

(Zawojewski, December 1986, pp. 20, 24–25; Ockenga and Duea, January 1984, pp. 28, 31)

92. Triangles, squares, rectangles, and trapezoids can result.
(Carroll, March 1988, pp. 6–11; Moses, December 1982, pp. 13–14)

93. 1; 6; 12; 8; 0.
(Bright, October 1978, pp. 28, 30–31)

94. *ABH, GHF, BHI, FHI, BCD, FED, BID, FID, BDF, BHF, BHD, FHD.*
(Ockenga and Duea, March 1979, pp. 28, 31–32; Greenes, February 1981, pp. 14–17; Devault, April 1981, pp. 40–43)

95. Two locations meet those conditions.

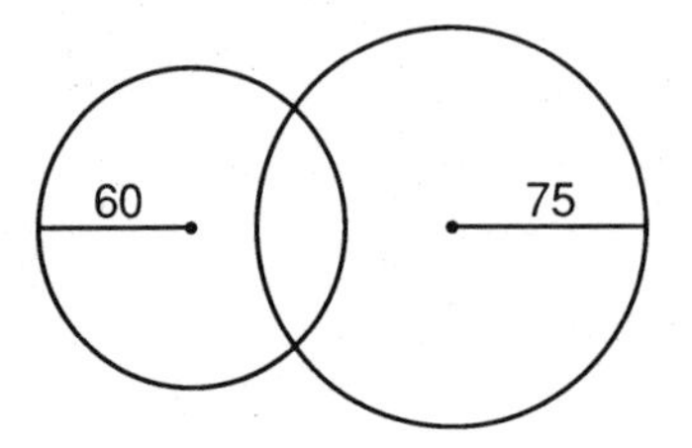

(Jamski, October 1982, pp. 28, 51)

96. (3,3), (6,6); (0,3), (3,0); (9,6), (6,9).

(Vissa, November 1987, pp. 6–10)

97. $p = 4s = 360'$, so $s = 90'$. Then $(2/3) \times (90') + 6'' = 60$ feet, 6 inches.

(Greenes and Immerzeel, January 1987, p. 26)

98. $c = 2\pi r = 2\pi(5/6)$ feet. 1 mile = 5280 feet. $5280/\left(\frac{10\pi}{6}\right) = 1008$.

(Ames, November 1977, pp. 50–53)

99. $\dfrac{2 \times 3.14 \times 93\ 000\ 000}{365 \times 24} = 67\ 000$ miles an hour.

(Vest, December 1981, pp. 32–33)

100. 93 square meters.

(Kroft and Yabe, October 1987, pp. 36–43)

101. No. The amount of area must remain the same. In this example, the seam down the middle is really a slight gap.

(Donegan and Pricken, May 1981, pp. 15–16)

102. ½ of (6)(14) + ½ of (6)(14) = 84 square meters, as does 6 × 14.

(Speer, November 1979, p. 23)

103. $A = s^2 - \pi^2 = (4.4)^2 - \pi\ (2.2)^2 = 19.36 - 15.20 = 4.16$ square centimeters.

(Teitelbaum, November 1978, p. 19)

104. Area of $\triangle AOB$ = ½ of (4)(4) = 8 square centimeters. Area of quadrant AOB = ¼ of πr^2 = ¼ of $\pi(4)^2$ = 4π square centimeters. Area of lune $= \frac{1}{2}(2\sqrt{2})^2\pi - (4\pi - 8) = 8$ square centimeters = area of $\triangle AOB$.

(Humphreys, February 1981, pp. 51, 62–64)

105.

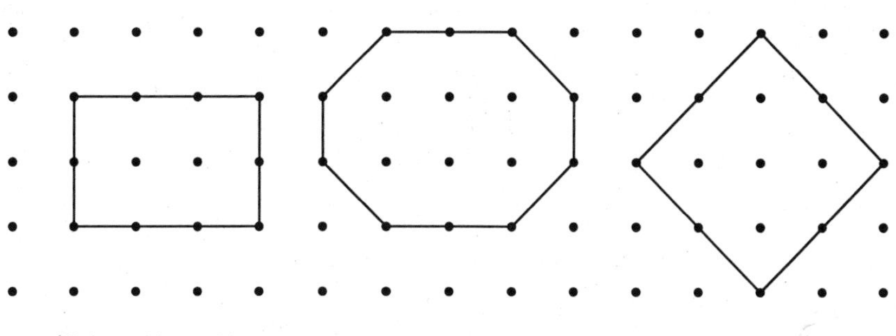

(Kolnowski and Okey, April 1987, pp. 26–33)

106. 5; 12; 6.

(Kolnowski and Okey, April 1987, pp. 26–33; Jamski, December 1978, p. 39; Leutzinger and Nelson, April 1980, pp. 8–9; Spitler, April 1982, pp. 36–38)

107. $A = \frac{x-2}{2} + y$.

(Litwiller and Duncan, April 1983, pp. 38–40)

108.

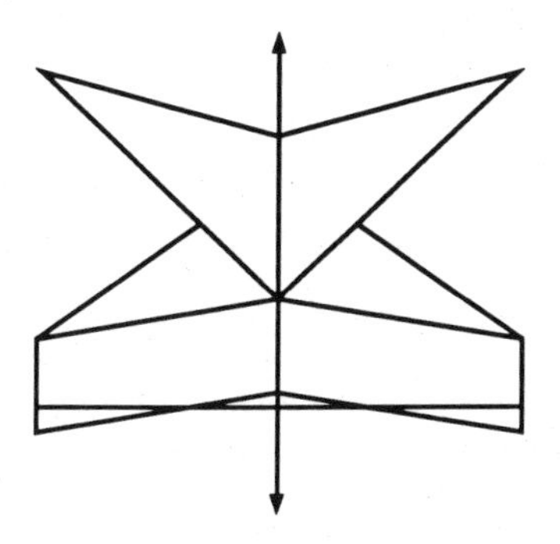

(Bazik and Turner, January 1983, pp. 28, 31)

109. A, B, C, D, E, H, I, M, O, T, U, V, W, X, Y. Expect some disagreement on proper lettering.

(Burns, December 1975, pp. 637, 641; Sanok, April 1978, pp. 36–40)

110. 24.

(Bidwell, March 1987, pp. 10–15; Jamski, December 1985, pp. 26, 48)

111. 102 units; 52 units; 202 units.

(Maier and Nelson, October 1986, pp. 34–35; Charles, February 1985, pp. 48–50)

112. Four; $2 \times 3 \times 2$ rectangular solid; $1 \times 1 \times 12$ rectangular solid.

(Maier and Nelson, November 1986, pp. 27–29)

113. 2×2 cutouts yield a volume of $(12 - 2(2)) \times (12 - 2(2)) \times 2 =$ 128 square centimeters.

(Burns, February 1976, p. 116)

114. 10.

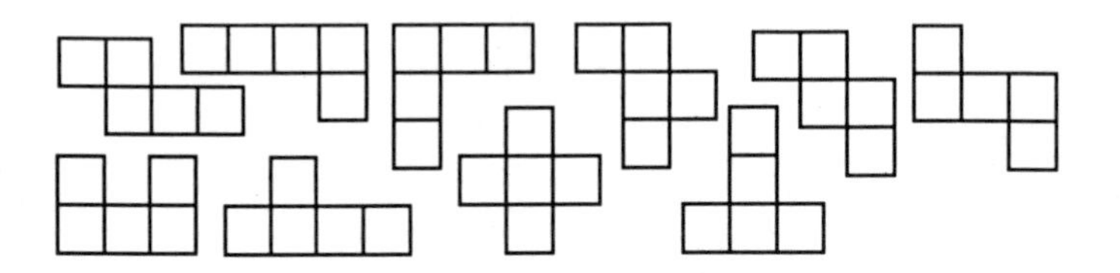

(Cowan, March 1977, pp. 188–90; Claus, January 1979, p. 57; Sanford et al., January 1979, p. 57)

115. 13.

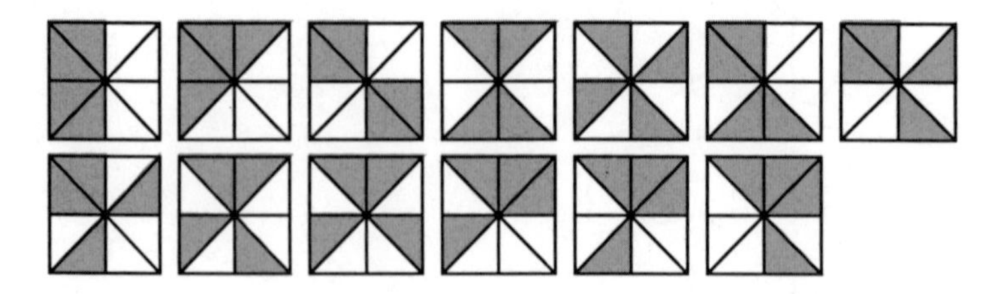

(Bruni and Silverman, April 1977, pp. 265–72; Lindquist, November 1979, p. 9)

116. 8.

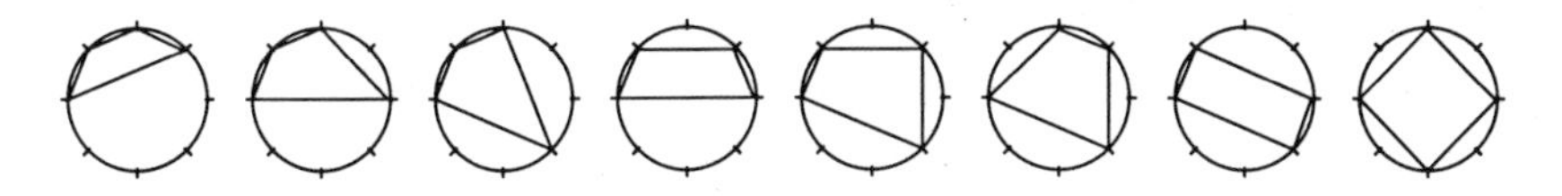

(Dana and Lindquist, January 1978, pp. 4–10; Aschenbrenner, January 1979, back of front cover)

117. radius; ½ of the circumference; ½ of $r(2\pi r) = \pi r^2$.

(Szetala and Owens, May 1986, pp. 12–17)

118. *For 8:*

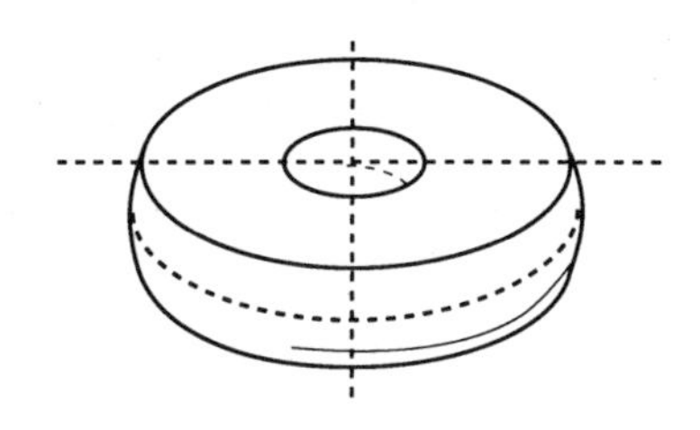

For 12:

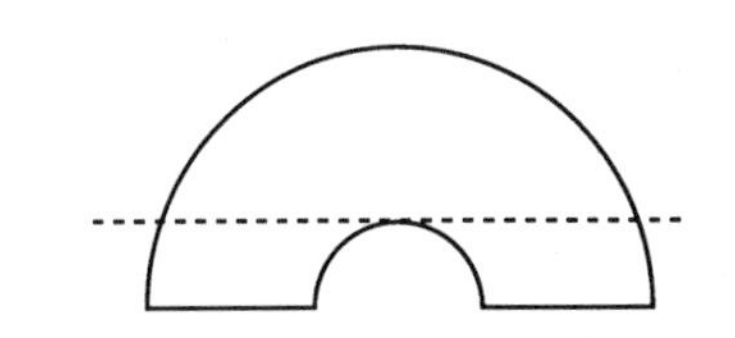

Take a horizontal cut as in previous solution, followed by a vertical cut through the center. Stack the four pieces and cut as pictured.

(Payne, May 1987, pp. 28, 32)

119. If n = the number of people, then $n(n - 1)/2 = 66$. $n = 12$.

(Charles, May 1986, pp. 26–27)

120. $(7 \times 6)/2 = 21$.

(O'Daffer, March 1985, p. 34; O'Daffer, December 1984, p. 30; LeBlanc, December 1977, pp. 16–20; Krulik and Rudnick, February 1982, pp. 42–46)

121. 2; 12; 7; ⅙.

(Fennell, March 1984, pp. 26–30; Shaw, February 1984, pp. 6–9)

122. $\frac{24}{x} = \frac{2}{10}$, $x = 120$.

(Vissa, March 1987, pp. 36–37; Souvinney, February 1986, pp. 56–57)

123. 55/84; 7/84.

(Wilmot and Barnes, March 1988, pp. 32–33)

124.

With five 3s	1
With four 3s	20
With three 3s	120
With two 3s	240
With one 3	120
With no 3s	0
	501

(Glatzer, September 1982, pp. 25, 30)

125.

1 6 15 20 15 6 1

1 7 21 35 35 21 7 1

(Jordan, December 1979, pp. 32–34; Stone, March 1980, pp. 47–49)